SURVIVAL

Louise McNaught

Text by Anna Claybourne

RP | KIDS
PHILADELPHIA

I would like to dedicate this book to my husband and soulmate, Rob, my amazing daughter, Autumn, my wonderfully supportive parents, and my brother, Pete—LM

To find out more about Tusk and
the wildlife projects they support,
visit www.tusk.org

Running Press Kids
Hachette Book Group
1290 Avenue of the Americas, New York, NY 10104
www.runningpress.com/rpkids
@RP_Kids

Printed in China

Originally published in hardcover September 2018 by Big Picture Press in the UK
First Running Press Edition: August 2019

Published by Running Press Kids, an imprint of Perseus Books, LLC, a subsidiary of Hachette Book Group, Inc.
The Running Press Kids name and logo is a trademark of the Hachette Book Group.

The Hachette Speakers Bureau provides a wide range of authors for speaking events.
To find out more, go to www.hachettespeakersbureau.com or call (866) 376-6591.

The publisher is not responsible for websites (or their content) that are not owned by the publisher.

This book features original artwork by Louise McNaught created under the following names:
Page 8 *Burned too Bright*; 10 *Unforgettable*; 12 *Fragmented Freedom*; 14 *Bananas*; 16 *True Blue*;
18 *From the Ashes*; 20 *Man of the Forest*; 22 *Born Slippy*; 24 *One Last Look*; 26 *Before Religion II*;
28 *Pandaemonium*; 30 *In the Red*; 32 *The Luzon Peacock Butterfly*; 34 *Paradise Lost*; 36 *Exhale*;
38 *Shellshocked*; 40 *Circle of Love*; 42 *FIN*; 44 *God's Last Lovesong*; 46 *Falling For You*

Map image vectorEps/Shutterstock.com
Made with paper from a sustainable forest
Consulted by Dr. Philip Wheeler
Edited by Carly Blake

Templar Publishing uses the Tusk logo under license from the Tusk Trust.
Tusk is not the manufacturer of this product.

Print book cover and interior design by Adam Allori.

Library of Congress Control Number: 2019930178

ISBNs: 978-0-7624-9636-5 (paperback), 978-0-7624-9637-2 (hardcover),
978-0-7624-9639-6 (ebook), 978-0-7624-9641-9 (ebook),
978-0-7624-9640-2 (ebook)

10 9 8 7 6 5 4 3 2 1

The publisher would like to thank Tusk, in particular Dan Bucknell and Charlie Mayhew,
for their valuable contribution to this book.

Publisher's Note: Conservation statuses and population information is taken from
the last available assessment information on the IUCN Red List™ (www.iucnrdlist.org)
at the time of publication.

CONTENTS

ABOUT THE ARTIST

"I became drawn to endangered species because I was feeling more and more strongly that I wanted to highlight the issue of conservation."

Louise has been illustrating animals for as long as she can remember. She completed a degree in Fine Art at the University of Greenwich in 2012, and she has continued to work as a professional artist ever since, exhibiting both at home in the United Kingdom and internationally. In 2016, Louise had her first solo show focusing on endangered species, called "Survival," at the George Thornton Gallery in Nottingham. Some of the art from that show is featured in this book.

When Louise depicts these beautiful creatures that we are in danger of losing forever, her aim is to draw the viewer's attention to the animals' presence and energy. She paints by hand in high detail, creating each hair with a brush stroke. Louise sets each animal against a brightly colored background to create a vivid contrast, bringing the creature into sharp focus. The upward drips and fading color of the animals hint at the delicate balance between nature and humans.

"Nature always has, and will forever be, my inspiration, and I hope my art inspires people to protect it before it's too late."—Louise McNaught

To find out more about the artist, visit **www.louisemcnaught.com**

FOREWORD FROM TUSK

The world is experiencing an extinction crisis. We are losing species between 1,000 and 10,000 times faster than the natural extinction rate, caused almost entirely by human activity.

It may not grab the headlines in the way poaching does, but habitat loss remains the greatest threat to wildlife and is the number one cause of decline for 85 percent of threatened species. This will only worsen as the human population continues to grow and consume ever more natural resources. Tusk works in Africa, where the human population is set to double by 2050 and quadruple by the end of the century—the fastest population growth in the world. Finding space for both people and wildlife to coexist is the ultimate conservation challenge, not just for Africa, but the world over.

All is not lost, at least not yet. All the amazing, unique, and incredibly special animals featured in this book are still with us. Some had been predicted to have gone extinct already, but still they cling on. Tusk has been involved in countless conservation successes through our partners and projects in Africa, and we know that animals can be pulled back from the brink of extinction if there is enough support to do so. We have seen it before with the black rhino, which appears in this book as its numbers plummet again.

We all have a fantastic opportunity to do more for wildlife right now. We must not wait too long: extinction is forever.

Charlie Mayhew, MBE
Co-founder and Chief Executive of Tusk

About Tusk

Tusk is a charity set up in 1990 to help protect African wildlife, including the African elephant, black rhino, and mountain gorilla.

Tusk helps local organizations make an even greater difference, and supports more than sixty field projects in more than twenty African countries. It not only works to protect wildlife, but also helps reduce poverty through sustainable development.

HRH the Duke of Cambridge became the charity's Royal Patron in 2005 and has been a proactive supporter ever since.

THE STORY OF CONSERVATION

Earth is home to millions of species of living things that can be found in every kind of habitat, from scorching deserts to freezing polar lands. New species are being discovered every day, but, sadly, many are being lost, too, as a result of human activity.

BIODIVERSITY

The huge variety of life, including animals, plants, fungi, and bacteria, is known as *biodiversity*. Biodiversity is important because every single species, or type, of living thing is connected to many others through different food chains. So, when species die out, or become extinct, their ecosystems—habitats and all the living things found there—become unbalanced and can be damaged.

Like all species, humans rely on the world's ecosystems being healthy. A balanced soil ecosystem means crops can grow, and a thriving ocean ecosystem means there are plenty of fish to catch.

Conservation is the practice of caring for living things and ecosystems to protect life on Earth now and in the future.

HUMAN IMPACT

Humans have only existed for a relatively short part of Earth's history—around 200,000 years, but in that time, we have had a huge impact. Our success at hunting, farming, mining, building, and expanding the human population has changed vast areas of our planet. Wild land has been made into farmland, cities, and roads; factories, vehicles, and waste have created pollution that has harmed ecosystems; and overhunting, overfishing, and habitat destruction has caused many species to go extinct.

THE MODERN CONSERVATION MOVEMENT

In the 1800s, growth in industry and the rapidly rising human population made these problems worse, and people soon realized that they should try to protect ecosystems and species for future generations. This was the start of the modern conservation movement, which has been growing ever since.

CONSERVATION STATUS

Founded in 1964, the IUCN Red List of Threatened Species™ has become the world's most comprehensive information source on the conservation status of species. Each species is allocated a status, according to its population and how likely it is to die out.

Find out more at **www.iucnredlist.org**

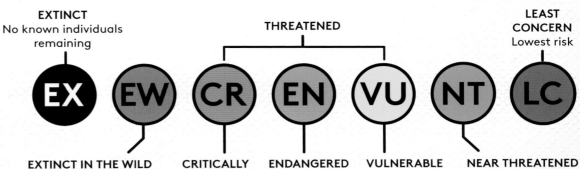

EXTINCT
No known individuals remaining

THREATENED

LEAST CONCERN
Lowest risk

EX EW CR EN VU NT LC

EXTINCT IN THE WILD
Known only to survive in captivity, or as an introduced population outside its original range

CRITICALLY ENDANGERED
Extremely high risk of extinction in the wild

ENDANGERED
High risk of extinction in the wild

VULNERABLE
High risk of endangerment in the wild

NEAR THREATENED
Likely to become endangered in the near future

CONSERVATION IN ACTION

With more and more awareness of wildlife conservation, there are many ways we are helping animals. Around the world, conservation operates on many levels, using a combination of laws, scientific studies, education, protected areas, and breeding programs.

EDUCATION
In the Arctic Circle, education on climate change and how it affects polar bears is helping encourage people to support conservation and live in more environmentally friendly ways.

ECOTOURISM
Responsible tourism driven by wildlife, such as whale watching off the northwestern United States, encourages countries to preserve species and wild areas.

REWILDING
Campaigns in Europe encourage people to keep gardens wild and bee-friendly.

CAPTIVE BREEDING PROGRAMS
Breeding species, such as pandas, in zoos and reserves can help increase their numbers and rebuild wild populations.

PROTECTED AREAS
National parks and nature reserves provide safe, wild areas of habitat for species. The Rana Terribilis Reserve in Colombia is helping protect the golden poison frog.

LAWS
Governments in Africa on have agreed international laws to ban trade in endangered species.

REDUCING CONFLICT
People are striving to live alongside animals with less conflict. In India, farmers grow crops elephants don't like, such as mint.

1930	1948	1960	1961	1962	1970s	1973	2006	2011	2012	2018
2 billion people	International Union for Conservation of Nature (IUCN) founded	3 billion people	World Wildlife Fund (WWF) founded	Rachel Carson's book *Silent Spring* raises awareness of damage caused by pollution	Extinction of the Caspian tiger due to habitat loss and hunting	Convention on International Trade in Endangered Species (CITES) established	The influencial Stern Review describes the effects of global warming	7 billion people	Extinction of the Pinta giant tortoise, after last of its kind, "Lonesome George," dies in captivity	Last male northern white rhino dies, leaving two surviving females

Aa	**LATIN NAME**	*Panthera tigris altaica*
EN	**STATUS**	Endangered
	POPULATION	Around 500, stable
	SIZE	8–10 ft from head to tail
	HABITAT	Temperate, often snowy forests
	LOCATION	Eastern Russia, with pockets in China and North Korea

SIBERIAN TIGER

The world's biggest cat, the Siberian or Amur tiger, is one of five surviving tiger subspecies. It lives the furthest north, mainly in eastern Russia, and it's well-adapted to this region's snowy forests, with large paws and thick fur. The Siberian tiger was hunted heavily for its fur and body parts such as bones and teeth in the early 1900s, and human conflict in their range during the Russian Civil War (1917–1922) also saw the loss of many tigers. By the 1940s, there were fewer than fifty left. In 1947, the former Soviet government stepped in and banned tiger hunting, becoming the first country to do so. Today, large-scale logging threatens to destroy tiger territories, which can be up to 186 sqare miles each (three times the size of Paris).

Thanks to a program of careful monitoring of tiger numbers and guarding against poaching, the Siberian tiger—though still rare and listed as Endangered—has come back from the brink of extinction, with a stable population of around five hundred as of the last census (population count) in 2010.

Aa	**LATIN NAME**	*Elephas maximus*
EN	**STATUS**	Endangered
	POPULATION	Around 50,000, decreasing
	SIZE	Up to 12 ft tall; up to 21 ft in length (including trunk)
	HABITAT	Forests, scrubland, and grassland
	LOCATION	Parts of Asia, from India in the east to Indonesia in the west

ASIAN ELEPHANT

Smaller than their African cousin, this species once roamed across much of Asia, with an estimated population of 100,000 in the early 1900s. Since then, numbers have dropped by half, with the biggest cause being habitat loss, as towns and farmland have replaced and broken up its forest home. With elephants and humans living so close, there is competition for food and space. Elephants increasingly raid crops, which farmers rely on for their livelihoods, and sometimes cause harm to people, resulting in human retaliation. Although a smaller problem in Asia than in Africa, illegal poaching for ivory, to be carved into ornaments, is still a threat to males.

Linking wild habitat areas with "wildlife corridors" and managing human-elephant conflict are key to helping this species. Planting crops that elephants dislike, such as mint, can help locals protect farmland. An international ban on the sale of ivory went into force in 1989, and more recently, conservation projects have supported a bigger number of local patrols to enforce anti-hunting laws.

Aa	**LATIN NAME**	*Equus zebra*
VU	**STATUS**	Vulnerable
	POPULATION	Around 25,000, increasing
	SIZE	6 ft 11 in.–8½ ft from nose to tail
	HABITAT	Grassy mountain slopes and meadows
	LOCATION	Southwestern tip of southern Africa

MOUNTAIN ZEBRA

The mountain zebra has two subspecies: the Cape mountain zebra and Hartmann's mountain zebra. Its natural home is hot, dry mountain grassland in southwestern Africa, and it is the highest-living zebra species, roaming to elevations of up to 6,561 feet. Unlike plains zebra, which form huge herds on the African savannah, mountain zebra live in small groups of just a few animals. In the early 1900s, sheep and cattle farming took over much of the mountain zebra's natural habitat. The zebra could no longer access their usual water sources, and they were forced out of farming areas to stop them competing with livestock for grazing ground. They were also hunted extensively for their skins and meat. A census in the 1930s revealed that the Cape mountain zebra was on the brink of extinction, with fewer than 100 left.

Setting up national parks has helped the species recover. The Mountain Zebra National Park in South Africa, founded in 1937, is now home to more than 700 Cape mountain zebra, and in 2008, its status was relisted from Endangered to Vulnerable.

Aa	**LATIN NAME**	*Gorilla beringei beringei*
CR	**STATUS**	Critically Endangered
	POPULATION	880, increasing
	SIZE	5–6 ft tall when standing
	HABITAT	High-altitude mountain forests
	LOCATION	Two populations in Africa, in parts of the Democratic Republic of Congo, Rwanda, and Uganda

MOUNTAIN GORILLA

The thick-furred mountain gorilla is a subspecies of the Eastern gorilla—one of two gorilla species—and is among our closest relatives. It lives in two locations in the mountain forests of Central Africa, at elevations of up to 13,123 feet. Though large and incredibly strong, these animals are peaceful plant eaters, living in family groups ten to twenty strong. Their natural habitat has been damaged by deforestation, which has forced them into smaller, higher areas. Diseases affecting humans, as well as human warfare, have also taken a toll on the mountain gorilla population, and poaching, sometimes for the exotic pet trade, continues to be a problem.

The mountain gorilla has been listed as Critically Endangered since 1996. However, conservation efforts have been working and they've seen a slow, steady increase in numbers. Most gorillas now live in protected reserves, and allowing tourists to visit them to see these great apes in the wild can encourage locals to protect the animals they live so close to and provide valuable paid work.

Aa	**LATIN NAME**	*Balaenoptera musculus*		**SIZE**	65–100 ft in length
EN	**STATUS**	Endangered		**HABITAT**	Oceans
⚇	**POPULATION**	10,000–25,000, increasing		**LOCATION**	Worldwide, apart from the Arctic and enclosed seas

BLUE WHALE

At the length of seven elephants, the blue whale is the largest animal ever to exist, and at the top of the food chain, it plays a vital role in the ocean ecosystem. Its sheer size and power was once no match for human hunters, and populations as big as 250,000 thrived. But, as whaling technology advanced in the late 1800s, it became a prime target for whaling ships, hunted for its meat, oil, and whalebone. The blue whale population fell heavily for nearly a century, to an estimated low of 5,000, until 1966 when a global hunting ban was imposed. Today, climate change is this whale's biggest threat. A single whale needs to eat 4 tons of shrimp-like krill a day, but global warming is melting Antarctic sea ice where krill are found.

Blue whales have one calf every two or three years, so it's no surprise that population recovery is slow. Ongoing conservation includes close monitoring of whale feeding, breeding, and migration zones to avoid contact with shipping lanes.

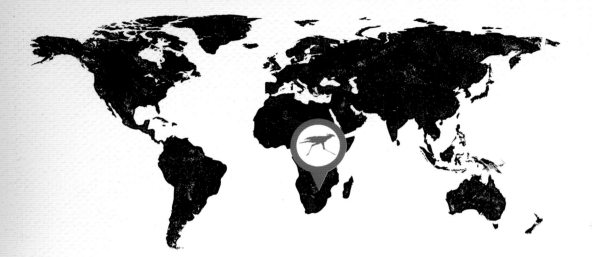

Aa	**LATIN NAME**	*Sagittarius serpentarius*
VU	**STATUS**	Vulnerable
	POPULATION	6,700–67,000, decreasing
	SIZE	3–4½ ft tall
	HABITAT	Open grassland, scrubland, and farmland
	LOCATION	Sub-Saharan Africa

SECRETARY BIRD

The secretary bird is a unique bird of prey. Unlike eagles or hawks, it seldom flies to hunt. Instead, this bird walks on the ground on its characteristically long legs, in search of snakes and rodents, and makes a kill by stamping on its prey with its powerful feet. A familiar sight across Africa south of the Sahara Desert, it's thought of as a common bird—but studies show its numbers have been falling rapidly since 2011, up to as much as 60 percent in some areas. This is probably due to a combination of urbanization, livestock farming, and agriculture encroaching on its habitat; collisions with cars and power cables; droughts (made worse by climate change); and capture for the exotic pet trade.

In their favor, many secretary birds already live in protected reserves, such as the Maasai Mara National Reserve in Kenya and Kruger National Park in South Africa. Outside of reserves, crop farmers welcome them because they hunt rodent pests. But careful monitoring is crucial to prevent this species from becoming Endangered.

	LATIN NAME	*Pongo pygmaeus*
Aa	LATIN NAME	*Pongo pygmaeus*
CR	STATUS	Critically Endangered
	POPULATION	About 100,000, decreasing
	SIZE	3½–4½ ft standing; arm span up to 8 ft
	HABITAT	Swampy and hilly tropical rain forests
	LOCATION	The island of Borneo in Southeast Asia

BORNEAN ORANGUTAN

The orangutan, which means "person of the forest," has three species—Bornean, Sumatran, and Tapanuli—and all are listed as Critically Endangered. Bornean orangutans have shorter beards, darker fur, and the males have large, distinctive cheek pads called *flanges*. All species play an important role in seed dispersal in the rain forests where they live. Since 1950, the population has dropped by more than half, due to extreme habitat loss—partly down to building, logging, and mining, but increasingly due to oil palm plantations. Since the 1990s, the demand for palm oil has exploded. It's used in more than half of all packaged products, from shampoo to makeup to snacks, and in Southeast Asia alone, plantations have already taken the place of tens of thousands of square miles of forest.

There are several reserves on the islands of Borneo and Sumatra, but many orangutans live outside of these. Conservationists work with loggers to keep patches of forest unharmed and with farmers to encourage sustainable palm oil farming.

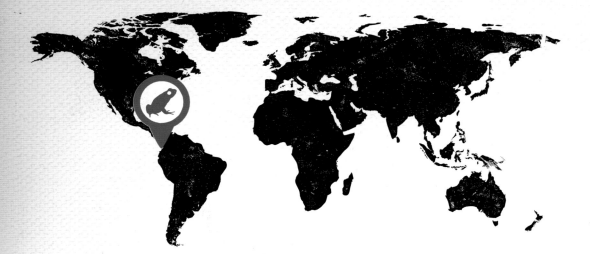

Aa	**LATIN NAME**	*Phyllobates terribilis*
EN	**STATUS**	Endangered
	POPULATION	Unknown (but abundant in its small home range), decreasing
	SIZE	1½–2 in. long
	HABITAT	Coastal tropical rain forests
	LOCATION	Small areas on Colombia's western Pacific coast

GOLDEN POISON FROG

Visit the jungles of Colombia's Pacific coast and you could well spot a golden poison frog. In the areas of forest where these small frogs are found, they are quite common. Unfortunately, though, their home range is relatively tiny, at just 3,106 sq miles, and it's threatened by deforestation, illegal logging, and gold mining, which has put this frog on the Endangered species list. Like all poison dart frogs, the golden poison frog produces a toxic secretion, and the local Emberá tribespeople use it to tip their blowpipe darts. Without harming the animal, they wipe their dart tips over the frog's back to gather the poison, and then use the darts to hunt small animals for food. The frogs' poison, which is one of the deadliest in the world, may also have a valuable medical use as a powerful painkiller.

The forests continue to shrink, but in 2012, a third of a mile of the golden poison frog's habitat was safeguarded with support from a conservation charity, creating the Rana Terribilis Amphibian Reserve—the first safe haven for this species.

Aa	**LATIN NAME**	*Panthera uncia*
VU	**STATUS**	Vulnerable
	POPULATION	4,500–6,500, estimated to be decreasing
	SIZE	5½–8 ft from head to tail tip
	HABITAT	Cold, rocky mountains
	LOCATION	The Himalayas and other central Asian mountain ranges

SNOW LEOPARD

High in the icy mountains of Central Asia lives the rare, elusive snow leopard. Its beautiful spotted fur is uniquely colored to camouflage this cat perfectly against its rocky backdrop, and extra-thick to withstand temperatures as low as –13°F. It is this coat that has made the snow leopard a target for poachers. Pelts (furred skins) can sell for over $1,000, which is more than a year's wages in many Central Asian countries. With more people setting up farmland in the snow leopard's natural habitat, these big cats have come to rely on farm livestock for part of their diet. However, this has resulted in farmers killing many of them to protect their herds. Snow leopard numbers are falling, but their harsh mountain home makes it difficult to monitor populations and guard them.

The twelve countries where snow leopards live, ranging from Russia to Myanmar, have agreed on a strategy that includes prosecuting poachers, helping farmers protect livestock, and maintaining the wild prey species of this big cat to help it survive.

Aa	**LATIN NAME**	*Diceros bicornis*
CR	**STATUS**	Critically Endangered
⚇	**POPULATION**	5,000, increasing
	SIZE	4½–6 ft tall; 9½–13 ft from head to tail
	HABITAT	Grasslands, forests, and wetlands
	LOCATION	Eastern and southern Africa

BLACK RHINO

Rhinos have been walking the Earth for millions of years, and today five species exist. The black rhino is one of two that live in Africa and one of three that have two horns instead of one. In the early 1900s, the black rhino population was estimated at around 100,000, but this had fallen to below 3,000 by 1995, and the species was declared Critically Endangered in 1996. This rapid decline was mainly due to poaching for rhino horn to be used in traditional medicine in parts of Asia. Even though there's no evidence it has any healing effect, the ground-up horn is worth more than its weight in gold. The horns are also sought after for making ornaments, jewelry, and even walking sticks.

Since the 1990s, black rhino numbers have gradually increased, and, by 2011, the population had doubled from its lowest point. This is thanks to protected reserves being set up and tireless conservation efforts that include animal monitoring, anti poaching patrols, and environmental education in the local community.

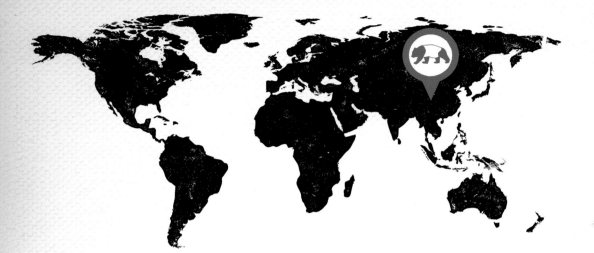

Aa	LATIN NAME	*Ailuropoda melanoleuca*
VU	STATUS	Vulnerable
	POPULATION	Around 1,800, increasing
	SIZE	4–5 ft from head to tail
	HABITAT	High-altitude broadleaf and mixed forests
	LOCATION	Central China

GIANT PANDA

This iconic black-and-white bear once lived across much of eastern and southern China and parts of Myanmar and Vietnam. Today, it survives in a handful of remote mountainous forests in central China. A panda's diet is almost entirely bamboo, and bears will feed for up to fourteen hours a day, eating twenty-two pounds or more of shoots. Further forest destruction and fragmentation are imminent threats, because the panda's habitat is located in the Yangtze River basin—China's economic center and home to more than half a billion people. The poaching of these bears for their uniquely colored fur also continues to impact the population.

Since the 1960s, the Chinese government has set up more than forty panda reserves. Conservationists are working on linking these unconnected locations with wildlife corridors so that pandas can meet and mate more easily and can move around to find bamboo. In 2016, thanks to successful captive breeding efforts, the giant panda's status was revised from Endangered to Vulnerable—a huge success.

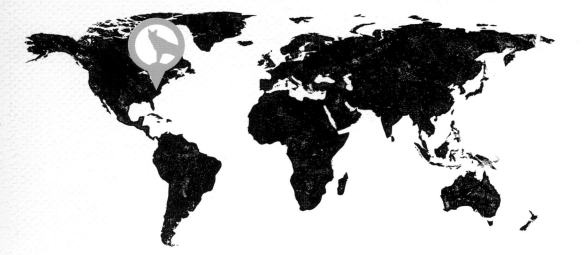

Aa	**LATIN NAME**	*Canis rufus*
CR	**STATUS**	Critically Endangered
	POPULATION	Less than 300, increasing
	SIZE	4½–5½ ft from nose to tail tip
	HABITAT	Forests, bushland, prairie (grassland), and swamps
	LOCATION	A small area of eastern North Carolina

RED WOLF

The tawny-furred red wolf is smaller than its cousin, the gray wolf, with longer legs, taller ears, and shorter fur. Before the 1500s, when Europeans arrived in North America, the red wolf (or *wa'ya*) featured prominently in Cherokee mythology, where it was regarded as being sacred. Its range once covered most of southeastern United States, but by the 1970s, more than 99 percent of its habitat had been lost or split into small areas due to widespread urbanization and large-scale agriculture. Seen as a danger to both farm animals and humans, red wolves had been hunted in large numbers, almost to extinction—just seventeen remained.

In 1973, a plan was drawn up to try to save the red wolf, and the U.S. Fish and Wildlife Service began a breeding program with fourteen of the wolves captured from the wild. This increased the population, and since the late 1980s, some wolves have gradually been released back into the wild in North Carolina, while others remain in captivity.

Aa	**LATIN NAME**	*Papilio chikae*
EN	**STATUS**	Endangered
	POPULATION	Unknown
	SIZE	4.3–4.7 in. wing tip to wing tip
	HABITAT	Mountain forests 4,921 ft elevation and above
	LOCATION	Luzon Island, Philippines

LUZON PEACOCK SWALLOWTAIL

The extremely rare Luzon peacock swallowtail is a large butterfly with shimmering, blue-tinged wings. It is endemic (only living in that place) to the island of Luzon in the Philippines, where it can be found high up in mountainous tropical forests. It was only discovered by scientists in 1965, and because of its large size and vibrant coloring, it quickly became a prized target for butterfly collectors, who pay up to $600 for a specimen. It also suffers from damage to areas of its habitat, as roads and tourism facilities continue to be built in the forests where it lives.

Unfortunately, the rarer the species becomes, the more collectors want to own one. Capturing and trading these butterflies is illegal, but laws can be hard to enforce. However, in 2017, three people were jailed for importing 2,800 dead butterflies from Southeast Asia, among which were Luzon peacock swallowtails. The Philippines is home to a number of other unique swallowtail species, which makes this a priority location for future butterfly conservation programs.

Aa	**LATIN NAME**	*Paradisaea rudolphi*
VU	**STATUS**	Vulnerable
	POPULATION	Up to 10,000, decreasing
	SIZE	11.8 in. from beak to tail tip
	HABITAT	Tropical mountain forests
	LOCATION	Papua New Guinea in Southeast Asia

BLUE BIRD-OF-PARADISE

Blue birds-of-paradise are magnificent-looking birds only found in Papua New Guinea in Southeast Asia. The male has spectacular blue plumage and two long tail streamers, which it displays during courtship dances to try to win a mate. However, this beauty makes the bird attractive to local hunters, who collect its colorful feathers to sell for traditional ceremonies and as tourist souvenirs. At the same time, large areas of forests where these birds live are shrinking, as trees are cleared to make way for homes and farming plots.

Although hunting the birds is already controlled by law in Papua New Guinea, it can be hard to police in remote areas. The creation of forest reserves, along with education programs in the community, is the best way to help this species survive—and this provides work for locals, allowing them to make a living from ecotourism. Conservationists aim to learn more about this species by monitoring the rate of forest loss, levels of hunting, and extent of trade in its home range.

Aa	**LATIN NAME**	*Ursus maritimus*
VU	**STATUS**	Vulnerable
⦿	**POPULATION**	20,000–25,000
〰	**SIZE**	5½–10 ft from head to tail
❦	**HABITAT**	Sea ice, islands, and coastal land areas
🌐	**LOCATION**	In and around the Arctic Circle

POLAR BEAR

Majestic, huge, and snow-white, the polar bear has never been common, due to the harshness of its freezing Arctic habitat at the top of the Earth. This species can be found in fewer than twenty separate populations across Canada, Greenland, Russia, and Alaska. The local Inuit people, who also inhabit the icy lands of the Arctic Circle, have hunted polar bears sustainably for thousands of years, for meat to eat and for their fur to make warm clothes. A few hundred years ago, this region remained sparsely populated by people and few bears were killed, but in the 1900s, more and more hunters were able to access the Arctic, using snowmobiles, icebreakers, and airplanes, and polar bear populations subsequently plummeted.

In 1973, the International Agreement on the Conservation of Polar Bears set strict limits for hunting, and the species began to recover—only to face another deadly threat: the warming of the oceans is melting large areas of Arctic sea ice, which polar bears rely on to hunt seals, their main food source.

Aa	**LATIN NAME**	*Eretmochelys imbricata*
CR	**STATUS**	Critically Endangered
	POPULATION	Fewer than 50,000, decreasing
	SIZE	Carapace up to 3 ft in length
	HABITAT	Coastal coral reefs, estuaries, and bays
	LOCATION	Tropical seas and oceans around the world

HAWKSBILL TURTLE

The hawksbill turtle gets its name for its characteristic birdlike beak. It lives in shallow, warm, coastal waters throughout the tropics, where it can find plenty of sponges, anemones, and jellyfish to eat, with sandy beaches for nesting close by. It's known for its colorful mottled shell, called a *carapace*, which has long been used to make hair decorations and jewelry. During the 1900s, hunting of these turtles for their shells led to their population plummeting by 80 percent. Other threats that have contributed to their decline include oil spills, entanglement in fishing gear, accidental capture, damage to coral reef habitats, and egg collecting.

The hawksbill is now protected under international and national laws, but this hasn't stamped out poaching, and illegal trade remains a big problem. Coastal nature reserves are extremely important, and beach cordons and regular patrols help provide safe nesting sites. Protecting beaches and coral reefs, as well as fisheries switching to turtle-safe fishing gear, are the key to saving this species.

Aa	**LATIN NAME**	*Lemur catta*
EN	**STATUS**	Endangered
⚬⚬	**POPULATION**	2,000–2,400, decreasing
〰	**SIZE**	15–18 in. long with a tail of 22–25 in.
🌿	**HABITAT**	Forests and scrubland
🌐	**LOCATION**	Southern Madagascar, Africa

RING-TAILED LEMUR

With its extraordinarily long, black-and-white ringed tail, this unique lemur is instantly recognizable. There are around one hundred species of lemurs, which are relatives of monkeys. All lemur species are only found in the wild on the African island of Madagascar and more than ten are listed as Endangered. The lemur's once-widespread natural habitat has been reduced to small, isolated forests by human activity, including the development of cities, the clearing of forests for farmland, and the felling of trees to make charcoal. This fragmentation makes it challenging for them to move around the forests safely to find food and mates.

Despite a hunting ban, lemurs—especially ring-taileds—are a target for pet-trade poachers, and an estimated 28,000 were taken from the wild between 2010 and 2013. Today, many national parks and reserves exist across Madagascar, providing safe habitats for lemurs and jobs for local people, while generating income through ecotourism that can support future conservation.

Aa	**LATIN NAME**	*Sphyrna lewini*
EN	**STATUS**	Endangered
	POPULATION	Unknown
	SIZE	10–13 ft in length
	HABITAT	Coastal areas, from shallow to deep water
	LOCATION	Tropical seas and oceans around the world

SCALLOPED HAMMERHEAD

The scalloped hammerhead belongs to a group of large iconic sharks with bizarre wing-like heads. It's one of many shark species facing endangerment due to overfishing, which has drastically reduced populations in the last century. Sharks are hunted for sport, for meat, and for their fins, which are used to make the Asian delicacy shark fin soup. Hammerheads, though solitary, sometimes gather in schools several hundred strong, and this behavior has made it easy for fishermen to take many sharks at one time. Sea creatures can be hard to monitor, and it is equally difficult to enforce fishing bans over vast oceans.

However, in 2014, the species became the first shark to be protected by the U.S. Endangered Species Act, one of the world's strongest conservation laws, and countries such as Brazil, Ecuador, and Mexico are gradually making stronger fishing laws to protect hammerheads. Several conservation groups campaign to discourage people from eating shark fin soup, and demand for the soup has declined.

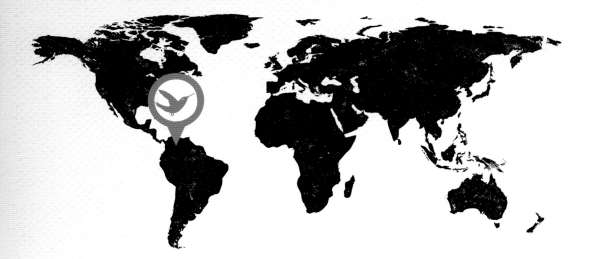

	LATIN NAME	*Lepidopyga lilliae*
	STATUS	Critically Endangered
	POPULATION	70–400, decreasing
	SIZE	3½ in. from beak to tail tip
	HABITAT	Mangrove forests
	LOCATION	Coastal areas of northern Colombia, South America

SAPPHIRE-BELLIED HUMMINGBIRD

The tiny, shimmering sapphire-bellied hummingbird is extremely rare. Very small in size, it has a range to match, being found only in a few areas of coastal mangrove forest in northern Colombia, which are shrinking fast. In the 1970s, a pipeline was built, which blocked and redirected the natural flow of the Magdalena River and the Caribbean Sea in and out of the mangrove swamp, killing off many trees in the hummingbird's natural habitat. Since then, urban development in the area has brought deforestation, industrial pollution, and sewage, further affecting the coastal swamp ecosystem and its wildlife.

Despite numerous searches in the area since 1990, there have been very few sightings. Although there are two national parks in the region, they have not always been able to prevent habitat loss. With fewer than four hundred birds remaining, the next steps are to restore and replant mangrove forests and enlarge the protected areas where the hummingbird lives in an effort to save it from extinction.

Aa	LATIN NAME	Genus *Bombus*
	STATUS	Various
	POPULATION	Decreasing
	SIZE	0.3–0.9 in. in length
	HABITAT	Gardens, meadows, farmland, forests
	LOCATION	Throughout the Northern Hemisphere, and several species in South America

BUMBLEBEES

There are about 250 species of bumblebees found throughout the northern hemisphere and South America. Like honeybees, they are important pollinators for plants, especially wild flowers and crops such as tomatoes and blueberries. Though some species are a common sight, a growing number are in serious decline. Several, including Franklin's bumblebee, which is only found in a 21,000 square mile area of the western United States, is listed as Critically Endangered. In modern farming methods, the loss of flower-filled hedgerows between fields in favor of fences makes it harder for bumblebees to find the flowers they feed on and suitable places to nest. The use of crop pesticides and diseases caught from honeybees have also caused bumblebee declines.

In 2011, the IUCN set up a group of seventy scientists to study bumblebees around the world. Conservation actions encourage farmers and gardeners to grow wild flowers and reduce pesticide use.

HOW YOU CAN HELP

Tusk is a charity set up in 1990 to help protect African wildlife, including the African elephant, rhino, cheetah, chimpanzee, mountain and lowland gorillas, African wild dog, and marine turtles. Tusk supports more than sixty field projects in more than twenty African countries that not only work to protect wildlife and its habitats, but also help reduce poverty through sustainable development.

☞ Visit www.tusk.org to learn about Tusk's wildlife projects.

☞ Make a donation to pledge your support. No matter how small, every little bit helps.

☞ Buy a gift from the Tusk Shop, at www.tusk.org/shop

REDUCE, REUSE, RECYCLE

Switch off lights and plugs when not in use. Give old clothes to charity shops or someone who needs them. Carry a refillable water bottle or cup instead of buying disposable ones. When waste is unavoidable, recycle as much of it as possible.

KEEP THE EARTH TIDY

Don't drop litter—always bin it. Only flush toilet paper down the toilet, not diapers, wet wipes, or cotton swabs.

SHOP SUSTAINABLY

Where you can, avoid items with packaging. Look out for locally produced foods that don't have to travel far to reach you.

DONATE TO A CHARITY

Hold a bake-off, a sports event, or another fun activity to raise money for a conservation charity of your choice.

VOLUNTEER

From local beach and park clean-ups to wildlife counts, you can volunteer to help an animal conservation project near you.

ADOPT AN ENDANGERED ANIMAL

Some charities and zoos allow you to adopt an animal for a small payment. In return, you get a certificate and regular updates on your animal.

TRAVEL RESPONSIBLY

Research before you travel. If you visit wildlife areas, at home or in other countries, make sure you know your trip will not cause harm to the habitat or the animals. Avoid buying souvenirs made from ivory, tortoiseshell, or other wildlife products.

BE A CONSERVATIONIST!

If you're passionate about conservation, you could work in this area as a career: working in wildlife parks, setting up conservation schemes, caring for animals, or campaigning for change. If you'd like to do this, choose science subjects at school, especially biology, zoology, botany, forestry, ecology, environmental studies, or conservation studies.